W9-DBT-999

INSTANT
Weather
FORECASTING

By the same author

Air Rider's Weather

Basic Windcraft

Cruising Weather

Dinghy and Boardsailing Weather

Instant Wind Forecasting

Reading the Weather: Modern Techniques for Yachtsmen

Sailing off the Beach

Weather Forecasting Ashore and Afloat

Wind and Sailing Boats

The Weather Handbook

The Wind Pilot

Published by Adlard Coles Nautical
an imprint of A & C Black Publishers Ltd
36 Soho Square, London W1D 3QY
www.adlardcoles.com

Text and photographs copyright © Alan Watts 1968, 2000, 2007, 2011

First edition published by Adlard Coles 1968
Reprinted 1968, 1971, 1973, 1975, 1976 (with amendments), 1979, 1981
Reissued in paperback by Adlard Coles Ltd 1985
Reprinted 1985, 1988, 1989 (twice)
Reprinted by Adlard Coles Nautical 1991, 1993, 1995
Second edition 2000
Reprinted 2001, 2003, 2004
Third edition 2007
Fourth edition 2011

ISBN 978-1-4081-3709-3

All rights reserved. No part of this publication may be reproduced in any form or by any means – graphic, electronic or mechanical, including photo-copying, recording, taping or information storage and retrieval systems – without the prior permission in writing of the publishers.

The right of the author to be identified as the author of this work has been asserted by him in accordance with the Copyright, Designs and Patents Act, 1988.

A CIP catalogue record for this book is available from the British Library.

This book is produced using paper that is made from wood grown in managed, sustainable forests. It is natural, renewable and recyclable. The logging and manufacturing processes conform to the environmental regulations of the country of origin.

Typeset in 8.5 on 9.5 Myriad
Design by Susan McIntyre
Printed and bound in Spain by GraphyCems

Note: while all reasonable care has been taken in the publication of this book, the publisher takes no responsibility for the use of the methods or products described in the book.

A 24-colour photograph guide to
weather forecasting from the clouds,
for use by walkers, farmers, fishermen,
yachtsmen, golfers, holidaymakers;
in fact anyone to whom the weather in
the near future is of vital importance.

INSTANT
Weather
FORECASTING

4th Edition

Alan Watts

ADLARD COLES NAUTICAL
London

AURORA PUBLIC LIBRARY

Contents

Preface to the Fourth Edition

In the hot summer of 1967 I reluctantly sat down to write *Instant Weather Forecasting*. It was an idea that I had recently tried out on James Moore, who was then Editor of the Adlard Coles imprint of Granada Publishing. He seemed more than keen that I should write this little book but I was busy putting together a physics textbook and I did not want to be deflected from, what seemed to me, this important work. I soon found that James was right—the physics textbook never saw the light of day but *Instant Weather Forecasting* has been in continuous print ever since its spectacular launch in 1968.

Up to that point I had only written two books with relatively small print runs, and I was staggered to find my publishers doing an initial printing of 75,000 copies so that they could supply the editions of foreign publishers across Europe as well as in America. Jim even felt confident enough to call a press conference to launch the book. I found myself writing captions to a bevy of my sky pictures for the prestigious *Sunday Times Magazine*. I found it amazing that my little book was heading the 'Hidden Best Sellers' list. It was a heady time for a young author.

The reviews were everywhere excellent. The reviewers seemed to respond to a book about the difficult subject of meteorology which they could immediately relate to, even if they didn't fully understand how I'd arrived at the inferences quoted in the book.

It is this chance to dive straight in and play the forecasting game as well as the attractiveness of the full-page pictures that explains why *Instant Weather Forecasting* has been so successful and is now being re-launched after forty three years. That small beginning has now developed into an extensive library of transparencies covering all aspects of the sky, as well as interesting phenomena like rainbows and sundogs, jack frost patterns and the fairyland that develops when freezing fog paints trees with hoar frost. However, such otherwise irresistably photogenic pictures are not for this book. *Instant Weather Forecasting* is an essentially practical book. It does not contain many of my most spectacular pictures simply because they do not properly illustrate the aspects of weather being discussed.

Instant Weather Forecasting cannot be used everywhere in the world. I have had quite a lot of correspondence about this over the years, and one letter from someone in southern Africa stated that the book didn't work. I had to reply that that was not surprising, as it was directed at countries in the temperate latitudes where depressions and anticyclones chase one another around the hemisphere, and no one could expect it to apply very much to the interior of Africa. So while *Instant Weather Forecasting* is good for anywhere between Vancouver and Vladivostok (or Santiago and Sydney) it loses its grip when we get into the tropics, particularly Africa and the Caribbean. And those who live in the shadow of great mountain barriers may find it not as efficient at telling the weather as it would be on the shores of Atlantic Europe.

Anyone who writes a book of fiction hopes it will become a bestseller overnight. Practical books like *Instant Weather Forecasting* take a little longer. I estimate that we've sold over a million copies in a dozen languages worldwide over the last forty three years. I hope that those who buy this new edition will be able to assuage their desire to become weather forecasters just as others have done before them.

Alan Watts
2011

How to use this book

General note to the reader

When the sky has a certain look about it then very often a certain sequence of weather follows. There is nothing certain about it however. We see watery skies and thundery skies, red skies and even blue skies, but can one be sure that the weather forecast written into the clouds, or lack of them, will turn out to be correct? Most of us know that sometimes a watery sky passes without a drop of rain, that thundery-looking clouds do not always bring thunder and that evening redness is not always followed by a quiet night.

In just twenty-four cloud studies there is only room for the most obvious inferences and sometimes the weather will not develop true to form. Even so there will be, perhaps, 75 per cent of occasions when the pictures in this book and the short notes which go with them will prove to be largely correct in foretelling the coming hours. So, to use the book, first find the sky which most resembles the one you have. Then see if the Major Clues are there as well. If they are, then read the Major Inference and the Explanation. If it still seems to fit the present situation consult the meteorological elements of greatest interest (say, Wind and Precipitation) to find the most likely trend and the normal period over which it will occur. An error in timing is the most likely mistake you will make. If bad weather is expected then it is better to assume it will come sooner than expected rather than later, especially if at sea.

To call any book on weather *Instant Forecasting* is, if taken literally, likely to give the impression that the road to becoming a forecaster of your own weather is easy and quick. This is not so. There are many years of experience in this book but others without professional forecasting experience have used the ideas put forward here and have found that they work. I hope they work for you.

When conditions are described without qualification then they tend to refer specifically to the temperate latitudes of the Northern Hemisphere. How to change the rules, which depend on wind directions and shifts, for the Southern Hemisphere is given on page 7. In general, a wind veer quoted in the text should be read as a back in the Southern Hemisphere and vice versa.

The message of the high clouds and the other phenomena from which short-term forecasts may be made are the same wherever the observer happens to be. It is as well to note, however, that in the lee of mountains or in the middle of continents or the ocean, wide divergences from some of the conditions described in this book are sometimes possible.

The temperature indications are based on 'sensation value' ie, what it feels like rather than just the actual air temperature. This is because wind speed is often of major importance in cooling the body and in so doing offsets the effects of high air temperture. Alternatively, a calm, clear winter's day may become bitterly cold without any change of air temperature just because the wind gets up. As most people wish to forecast the weather for recreation and holidays, so the major inferences given are generally for those months of spring, summer and autumn when the weather is fit for outdoor activity.

A gap in the brief statements of the likely trends and changes merely indicates that nothing of great use can be said under these headings. It does not indicate that there will be no changes at all, but that any that do occur will be relatively minor.

The Crossed Winds Rules

Changeable, and often unpleasant, weather comes about when a succession of lows (or depressions) passes over any locality. Surface winds tend to back (move anti-clockwise) when poorer weather is coming and to veer (move clockwise) when better weather is on the way. These are, however, rather vague rules for on the extreme right of Fig 1 it will be seen that the wind is backing as the fair weather comes along. Also, wind veers to the south of the tracks of depression but backs to the north as the lows pass. **NB** In the temperate latitudes of the Southern Hemisphere read *back* for *veer* and *veer* for *back*.

Much more useful rules are illustrated in Fig 1 where a simplified surface depression with its fronts is moving eastwards under the impetus of high speed upper winds. For our purposes the wind which flows parallel to the isobars near the surface is called the lower wind (L), and the wind at the level of medium and high clouds is called the upper wind (U).

These winds are not independent of one another—the way they flow across or parallel to one another is governed by the temperature of the major air block which is coming along. Thus, at (A), where the warm air mass of the depression is going to arrive later, the lower wind is crossed at right angles with the upper wind. It is, equally, crossed at right angles, but in the opposite sense, at (C) behind the cold front, where the really cold air mass is coming in. This will eventually change from blustery showers to fair weather.

At (A) the weather is going to deteriorate, while at (C) it is going to improve. Thus we arrive at the following rules which we will be calling the *Crossed Winds Rules*.

1. Stand with your back to the lower wind and if upper winds (or clouds) come from the left hand then the weather will normally deteriorate.
2. Stand with your back to the lower wind and if upper winds (or clouds) come from the right hand the weather will normally improve.

These rules apply to the Northern Hemisphere. They are applicable to the Southern Hemisphere (temperate latitudes) if you stand facing the lower wind.

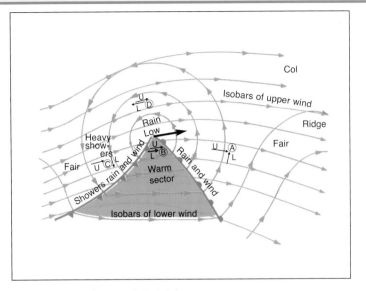

Fig 1 *To illustrate the Crossed Winds Rules.*

Strictly speaking, the lower wind is not the surface wind. Ideally the lower wind direction is the direction of motion of cumulus, stratocumulus or other low clouds. The surface wind blows at an angle to the isobars out of high pressure in towards low pressure. The angle is an average of 30° over land and 10° over the sea. So to find (L) from the surface wind (making sure it is not a land or sea breeze or a mountain or valley wind) stand back to the surface wind and rotate clockwise through the requisite angle. You will then be standing roughly back to the lower wind.

The third part of the rules is illustrated at (B) and (D) in Fig 1. Here the air masses are, in fact, all the same air mass from top to bottom of the troposphere and the upper and lower winds are flowing either parallel in the same direction or parallel in opposite directions. This means that no major front is in the immediate offing and so we can say:

3. Stand with your back to the lower wind and if upper winds (or clouds) move on a parallel course the weather will normally not change very much for a while.

Hints on recognising the direction and speed of the upper wind from the clouds are given in the descriptions accompanying the photographs, which have, in many cases, been chosen to help in using the rules for forecasting. They are not rules for forecasting tomorrow's weather (although they may help). They are for the next few hours or half a day or sometimes more, depending on what sort of cloud is in the sky when the forecast is made.

This description of the wind orientation rules does not purport to be complete in every aspect, but is ample for the purposes of this book. Readers who would like to delve further into the subject may like to refer to my book *The Weather Handbook* which will provide much more detail.

The numbers below refer to the skies depicted in the book.

Air Masses come from semi-permanent source regions which are large anticyclones. The following are those of the North Atlantic area.

Name	Abbreviation	Typical weather	Source
Maritime Tropical	mT	Extensively cloudy with rain and drizzle. Poor visibility and fog. 9, 10, 22	Azores High
Maritime Polar	mP	Showers and bright periods. Good visibility. 11, 12, 13, 15, 24	Polar High
Returning Maritime Polar	rmP	Cool but fair. Good visibility. 19, 21	As mP but modified by Atlantic Ocean
Continental Polar	cP	Intensely cold and often cloudy in winter	Siberian High
Continental Tropical	cT	Very warm and often cloudless. 7, 17	Southern Europe or North Africa

Anticyclone or High A region of mainly subsiding air with outflowing winds and often clearing skies. Sometimes, especially in winter, anticyclones are totally overcast for days on end.

Castellanus is normally lines of altostratus cloud out of which grow turrets of cumuliform clouds. It is a cloud which presages thundery outbreaks some hours hence. See 7 and 17.

Cloud Cover is measured in eighths of the sky covered. In our usage it will tend to mean the cover which is likely to affect the area from where the observations are made. For example, a western sky full of cirrus and

Explanation of terms used in the text

altostratus from an advancing front will constitute half cover, ie, four-eighths or so, but nondescript clouds on the horizon of an otherwise clear sky and which are obviously not advancing over-rapidly need not be counted as cloud cover at all.

Cloudy A cloudy sky is one where the total cloud wholly covers or nearly covers the entire sky.

Col The slack pressure region between two highs or two lows.

Cold Front The zone of division between a warm air mass (usually tropical maritime) and a cold one (usually polar maritime). Cold fronts tend to pass in half the time that warm fronts take to pass. The cold frontal surface slopes backwards and has about twice the slope of the warm front. (Fig 2, page 10).

Cyclonic Winds vary cyclonically if they change progressively in accordance with the way that is expected when a depression (or cyclone) moves across the observer. Winds which are cyclonic move progressively clockwise as a low centre moves to the north of the observer in the Northern Hemisphere but they move anti-clockwise when it passes to the south.

Depression or Low A region of mainly ascending air accompanied by precipitation and strong winds. A low is a bad weather system; old lows though, without fronts, may not bring very bad weather.

Diurnal Variation Air temperature near the earth's surface is at a maximum soon after midday (sun time) and at a minimum soon after dawn. The change from maximum to minimum and back again is called diurnal variation of temperature. Wind speed and cumulus cloud cover follows the same diurnal variation, being at its maximum in the early afternoon and at its minimum around dawn. Humidity is exactly opposite, being lowest in the afternoon and highest around dawn.

Fallstreaks are 'showers' of ice from high clouds—usually cirrus—which sink through many thousands of feet of the upper atmosphere. As they sink they become deflected by falling into wind at lower levels which is normally less strong than that above when ahead of bad weather systems. They, thus, appear to trail backwards from the heads of the parent cloud from which they originate. Because of the interrelation of

winds and temperatures of air masses the general direction in which the tails of the streaks point is towards a warm air mass. Thus, fall-streaks ahead of warm fronts often point to the SW, and those behind cold fronts point to the SE. (NW and NE, respectively, in the Southern Hemisphere.)

Dense streaks combined with a very strong drop in wind speed as they fall produce the jet stream banners of **1**. They produce streaks in one general direction ahead of less strong bad weather systems as in **2** and **3**, and behind such systems as described in **16**. They are dense, twisted and much more vertical ahead of thundery troughs in summer. When they sink in several directions they mean that good weather will persist. If they stream out parallel to the surface wind then they indicate no immediate change, as in **20**.

Fine and Fair In this book the term fine means that the skies are mainly clear, with less than two to three-eighths cloud cover, while fair means a partly cloudy sky with between three and six-eighths cloud cover. In either case there is no precipitation. At night the word clear has the same meaning as fine.

Floccus A form of altocumulus cloud associated with thundery weather, which resembles a flock of woolly sheep. Often seen with castellanus.

Fog may come from radiation cooling the ground **14** by being moved in on the back of a wind (advection fog), **10** or when a warm wind arrives over cold ground, as during a rapid thaw. Risk of radiation fog increases with the night in quiet conditions, and is aided by industrial and domestic smoke. Places downwind or downhill from built-up areas are particularly prone. Radiation fogs normally disperse with the sun, but encroachment of cloud may cut off the sun and make for a persistent fog. If in winter this cloud is stratocumulus then smog may form.

Advection fogs come in ahead of encroaching fronts and also when sea breezes start to blow on spring and early summer days. Increase of wind sometimes lifts fog into low stratus cloud.

Fractus is a form of stratus and is associated with the ragged fragments seen when this low cloud is forming or dispersing.

Explanation of terms used in the text

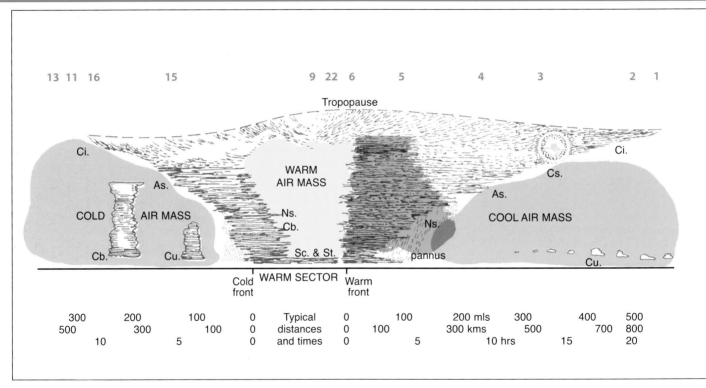

Fig 2 *Cross-section through the front of a developed depression. The distances and times are typical ones and are given for clouds and weather ahead of the warm front and behind the cold front. The warm sector may last for many hours or may be entirely absent. In the latter case the depression is occluding. The resultant occluded front is then formed by moving the cold and warm fronts into coincidence. Figures in grey above the diagram show where the skies with these numbers might be seen.*

Explanation of terms used in the text

Frost needs cool air and clear skies. It will form first in hollows and low-lying areas protected from the wind. The ground will fall below freezing first (ground frost). If the cooling continues it will extend slowly into the air deck above the cold ground (air frost). If a frost occurs with a moderate wind then this is likely to be a severe frost.

Inversions The air temperature normally falls with height at an average 'lapse rate' of 4.5°F per 1,000 ft of ascent (8°C per km). When, as at night in quiet conditions, the earth cools, then the air over it cools, leaving warmer strata above. Then the lapse rate is reversed and temperature increases for a certain distance with height. This deck of air where the lapse rate is reversed is called an inversion. Very strong inversions occur when air subsides from higher levels and warms up. The tropopause is a strong and permanent inversion which seals all the weather processes below it. Inversions at any height tend to do the same. They trap layer clouds like Sc and As.

Jet Streams High speed rivers of wind snaking round the temperate and sub-tropical latitudes of both Hemispheres. They are normally at about 30,000 ft (10 km) and are associated with the formation and movement of depressions. Characteristic clouds 1 form with them and so help to foretell the coming bad weather. Unseasonal weather occurs when the jets move from their normal latitudes.

Knot A speed of one nautical mile per hour. The normal unit for wind speed throughout the world. 1 kt = 1.15 mph = 1.855 km per hr.

Nebulosis is the form of stratus associated with total cover at a very low level 22.

Occluded Front Cold fronts move faster than warm ones so that eventually a cold front overtakes a warm one and the warm air becomes progressively lifted from the surface. In Fig 3 (page 13) this process has already started in the low near New York and it is on the point of starting in the low over the Faeroes. The zone of division between cool air ahead and cold air behind is called a cold occlusion. If the air behind the front is warmer than ahead it is a warm occlusion. Most land areas experience more occlusions than other types of front.

Orographic Rain Rain formed by the lifting of a moist airstream over hills. Over mountains the precipitation may be snow or sleet at any time of year. Airstreams which produce no precipitation over open country or coastal plains may well produce heavy and continuous rain, showers, thunderstorms etc when forced to traverse hill ridges.

Pannus is a form of stratus which is caused by turbulent wind eddies lifting and cooling the air-space below a cloud base from which rain, snow or sleet is falling. The latter will not necessarily reach the ground, as it evaporates as it falls and so moistens the air-space. Pannus foretells immediate rain, but once it has begun to rain in earnest it may extend into total cover at 1,000 ft (300 m) or so. See 5.

Pressure Tendency The fall of pressure with time (ie, tendency) is taken here to mean the following:

Major fall (or rise)	8–10 millibars or more in 3 hours
Rapid fall (or rise)	6–8 mb/3 hrs
Moderate fall (or rise)	between 3 and 6 mb/3 hrs
Slow fall (or rise)	less than 3 mb/3 hrs

The tendency is taken over 3 hours to allow time for a trend to reveal itself as a real one and not just a temporary small change. (Note that in some countries the pressure unit of hectopascal hP is used, however the hP is exactly the same change of pressure as the millibar.)

Ridge of High Pressure A region, usually the extension of a high where air is subsiding. See Fig 1.

Seabreezes Local winds that blow onto coasts during the days of spring and summer. They need light morning wind and sunshine, may penetrate tens of miles inland, and may draw their air from just as far seaward.

Seabreeze front When light winds from the land try to oppose seabreezes, a line like a miniature cold front develops between the two winds. Here the winds change from blowing off the land to blowing on to it.

Showers grow in unstable airstreams which are moist (eg maritime Polar air). There may be little diurnal variation over the sea or on windward facing coasts, but inland showers rarely spread themselves evenly over the day. Hill slopes force showery airstreams to disgorge their moisture as rain or snow and so shower-free areas may exist in the lee of hills and mountain ranges. Inland, mornings that are clear and cool may break out into showers by the late morning or early afternoon **11**. Air mass troughs **12** often reserve most of the showers for themselves. Nights are clear inland away from the windward coasts and stars are very bright.

Stable Airstream An airstream is stable if air which is forced to ascend in it, either by thermals or by being lifted over hills or frontal surfaces, tends to sink back again. An airstream may have stable layers over unstable ones or vice versa. A temperature inversion separates a lower, unstable layer from a stable one higher up. The signs of stable airstreams are layer clouds, sinking funnel or chimney smoke which tends to hang together and is not easily dispersed, poor visibility and extremes of temperature. Airstreams which are warmer than the surface over which they travel are stable near the surface.

Stratosphere The region of the upper atmosphere above the tropopause. No weather processes of any significance occur there.

Summer Half of the Year That period of the year which comprises the warmest and sunniest months with least cold weather and least major storms.

Thunderstorms may be single (air mass thunderstorm) **13**, in lines (frontal thunderstorms) **8**, or in areas often associated with slack pressure (col). A single storm is a cell of rising and sinking air currents with an active lifetime of 20 minutes to half an hour. When aided by frontal ascent (particularly cold fronts) then downdraughts due to rain and hail produce cold 'whalebacks' of air around a cell and lift the air there into new cells. These daughter cells take over from the dying parent. Medium-level storms have their bases some 6,000–10,000 ft (2,000–3,000 m) aloft and may follow the skies shown in **7** and **17**, depending on the winds aloft and below.

Thundery Rain follows the appearance of chaotic streaks, dapples, lines and waves in the build-up of the medium cloud ahead of a front. It is heavy here and there but thunder does not necessarily occur—just thunder-type spots.

Thundery Sky has cloud elements arranged in lines but the different directions of alignment at different levels give the appearance of chaos. It need not be entirely associated with the summer half of the year. The Crossed Winds Rules can be used to differentiate the forecast of **7** from that of **17**.

Trails formed by aircraft help to tell wind speed and direction aloft. When they are dense they indicate the possible incidence of a warm front. When they do not persist they indicate a relatively dry upper atmosphere and therefore little chance of major deterioration. High speed upper winds shred trails rapidly. If they shred sideways then the wind is across the trails, as in **3**. If they form castellated tops then the wind is parallel to the trails.

Tropopause The region of the atmosphere above about 8 miles (13 km) up which puts a 'lid' on weather processes by preventing any further ascent of air from the troposphere. It is characterised by temperature remaining constant with height.

Troposphere The region of the atmosphere in which weather processes occur. It is characterised by the temperature falling off with height, a situation which is arrested by the tropopause (see *Unstable Airstream* and *Inversions*).

Trough A region, often the extension of a low, where air is mainly ascending, leading to rain, showers etc, and temporary deterioration. There are two types of trough—1. frontal and 2. air mass. The former owes its existence to a front (usually an old one), while the latter forms in unstable polar airstreams especially in the rear of depressions.

Unstable Airstream An airstream in which ascending air continues to do so for a considerable depth of the atmosphere. An inversion will eventually stop the ascent and cut off the clouds at a certain upper level. The highest (and most impenetrable) level that clouds can reach is the tropopause. Cirrus and cumulonimbus must both have their tops below the tropopause which is about 8 miles (13 km) aloft over temperate latitudes of either Hemisphere. It is about 5 miles (8 km) aloft over the Poles and 10 miles (16 km) up over the Equator.

Explanation of terms used in the text

Unstable airstreams are characterised by cumuliform clouds lifting and dispersing smoke, good visibility and moderate temperature.

Veering and Backing The wind is said to veer when it changes direction clockwise and to back when it changes direction anti-clockwise.

Virga is trailing wisps of low cloud seen under the lowest parts of active fronts as they are passing. It may nearly reach the ground, especially when it snows.

Warm Front The zone of division between a cool or cold air mass and a warm one. The representative air masses are maritime polar ahead of the front (often warmed up and stabilised by time) and tropical maritime air behind.

The warm front is where a warm frontal surface meets the earth's surface, and the air behind the front gives its name to the front.

Warm Sector The block of tropical maritime air enclosed by the warm and cold fronts of a depression (Fig 3).

Winter Half of the Year That period of the year in any latitude which comprises the worst months, ie coldest and windiest with most major storms. It need not coincide with the official dates of winter and can extend from autumn through to spring.

Fig 3 *This actual weather map of part of the Northern Hemisphere illustrates possible places where the skies 1 to 24 might be seen. The thick black arrow is a jet stream. It is only part of the total jet which will snake more-or-less continuously round the temperate latitudes. Jet stream cirrus, depicted as appearing at the head of the jet 1 may be seen anywhere just to the south of the jet but is often obscured by the cloud of the depression. Cross-section through the fronts of a depression in Fig 2 could be either along say 55°N spanning Britain, or along 35°N across the warm sector of the depression centred round New York.*

Facts about clouds

Clouds can be divided into three height decks:

1. The region of low clouds—cloud bases between land surface and 7,000 ft (0–2 km).
2. The region of medium clouds—cloud bases between 7,000 and 25,000 ft (2–8 km).
3. The region of high clouds—cloud bases between 16,000 and 45,000 ft (5–13 km).

These divisions are generally applicable, but medium and high clouds sink below their lower limits in winter and with increasing latitude. In the tropics all clouds tend to rise above the upper limits for their type. Thus the divisions are only a guide. The best indication is the cloud type, eg cirrus is a higher cloud than altostratus, and altostratus higher than cumulus, etc.

Clouds are further classified by shape:

a Heap clouds (cumuliform)
b Layer clouds (stratiform)
c Feathery clouds (cirriform)

Having established the general cloud divisions the following table gives the major types. Height and shape numbers refer to the divisions given above.

Height and shape	Name	Abbreviation	Description
1a	Cumulus	Cu	Low heap clouds of small vertical extent
1ab	Stratocumulus	Sc	Low heap-layer clouds either broken or covering the whole sky
123a	Cumulonimbus	Cb	Heap clouds of great vertical extent producing showers and thunderstorms
1b	Stratus	St	Low, or very low, amorphous layer clouds covering coasts and hills
123b	Nimbostratus	Ns	Layered—often solid—clouds of great vertical extent associated with bad weather
2ab	Altocumulus	Ac	Medium level heap clouds in 'islands' or 'rafts' seldom covering the whole sky. The normal form is often associated with the term 'mackerel sky' but other forms are lenticularis (lens-shaped) and castellanus and floccus (chaotic clouds of thundery skies)
2b	Altostratus	As	Medium level layer clouds associated with the term 'a watery sun'
3c	Cirrus	Ci	High clouds mainly formed by more-or-less dense heads from which fall streaks or trails (fallstreaks) so producing hooks, banners, etc. Associated with the term 'mares' tails'
3ab	Cirrocumulus	Cc	High heap clouds ranged together in sheets of dappled, rippled etc. appearance. Also associated with the term 'mackerel sky'
3b	Cirrostratus	Cs	High amorphous cloud often covering the entire sky and sometimes only revealed by the presence of haloes about sun or moon.

Beaufort Scale of wind force

Beaufort number	General description	Sea criterion	Landsman's criterion	Velocity in knots
0	Calm	Sea like a mirror	Calm. Smoke rises vertically	Less than 1
1	Light air	Ripples with the appearance of scales are formed but without foam crests	Direction of wind shown by smoke drift but not by wind vanes	1 to 3
2	Light breeze	Small wavelets, still short but more pronounced. Crests have a glassy appearance and do not break	Wind felt on face. Leaves rustle. Ordinary vane moved by wind	4 to 6
3	Gentle breeze	Large wavelets. Crests begin to break. Foam of glassy appearance. Perhaps scattered white horses	Leaves and small twigs in constant motion. Wind extends light flags	7 to 10
4	Moderate breeze	Small waves becoming longer. Fairly frequent white horses	Raises dust and loose paper. Small branches are moved	11 to 16
5	Fresh breeze	Moderate waves, taking a more pronounced long form. Many white horses are formed. Chance of some spray	Small trees in leaf begin to sway. Crested wavelets form on inland waters	17 to 21
6	Strong breeze	Large waves begin to form. The white foam crests are more extensive everywhere. Probably some spray	Large branches in motion. Whistling in telegraph wires. Umbrellas used with difficulty	22 to 27
7	Near gale	Sea heaps up and white foam from breaking waves begins to be blown in streaks along the direction of the wind	Whole trees in motion. Inconvenience felt when walking against the wind	28 to 33
8	Gale	Moderately high waves of greater length. Edges of crests begin to break into spindrift. The foam is blown in well-marked streaks along the direction of the wind	Breaks twigs off trees. Generally impedes progress	34 to 40
9	Strong gale	High waves. Dense streaks of foam along the direction of the wind. Crests of waves begin to topple, tumble and roll over. Spray may affect visibility	Slight structural damage occurs (chimney pots and slates removed)	41 to 47
10	Storm	Very high waves with long overhanging crests. Foam in great patches is blown in dense white streaks along the direction of the wind. The surface takes on a white appearance. The tumbling of the sea becomes heavy and shock-like. Visibility affected	Seldom experienced inland. Trees uprooted. Considerable structural damage occurs	48 to 55

Jet stream cirrus

MAJOR INFERENCE A vigorous cyclonic situation exists upwind and so gales may blow up within the next 8–15 hours.

MAJOR CLUES 1. Long parallel banners formed by cirrus heads showering ice crystals into slower wind below. When strongly developed the fallstreaks combine into bands stretched along the direction of the upper winds. This direction is normally west to east or north-west to south-east (as in the photograph). (South-west to north-east in the Southern Hemisphere.) 2. Cirrus elements actually seen to move across the sky are usually going at over 80 knots (94 mph) and marked movement indicates real jet-stream speeds of 100–160 kt.

EXPLANATION Jet streams play a major role in the formation of young and vigorous surface depressions. These develop some hundreds of miles on the equatorial side of the jet axes. So do the characteristic jet stream clouds and, thus, the latter show that a vigorous depression may well be forming and moving in soon. But as the banners are dense when fully formed, they take some time to disperse and so can outlive the growth phase of the depression with which they formed. However, the Crossed Winds Rules will help in assessing the likelihood of real deterioration. The aircraft trail which is spreading sideways confirms that the wind up there is from bottom right, ie in the direction of the banners.

Element	Trend	Normal change	Normally in	Risk of	Possibly
Wind	Increase and back into southern quadrants	Force 2–4 W to NW becoming Force 5–8 SW to SE	6–12 hours	Force 6 in 3–6 hours	Severe gale or storm later, esp at sea in winter half of the year
Visibility	Good or exceptional then decrease	30 mls or more at first, decreasing to 10 mls or less	10–15 hours	Sea and hill fog by end of period	
Precipitation	Rain or snow	Continuous, moderate later	10–15 hours	Heavy rain or snow later	Light or intermittent if situation does not develop
Cloud	2–4/8 Ci, increasing from west	8/8 Cs, then 8/8 As, 8/8 Ns, Sc and pannus when real deterioration	10–15 hours	6–8/8 Sc forming to obscure upper cloud build-up	3–6/8 high and medium cloud if situation does not develop
Temperature	Becoming cooler	Feels cooler as sun is cut off and wind increases	5–10 hours	Bitter in winter	Little change if wind does not develop
Pressure	Falling	Moderate fall under Ci but speeding up. Rapid fall leads to Force 6–8 but major fall may give Force 9–10 at sea	12–24 hours	Major fall and more rapid (halve the times) passage of depression	Slow fall and so longer build-up of wind etc

2 Cirrus and cirrostratus

SKY WHICH MEANS DETERIORATION

MAJOR INFERENCE The warm front of a depression is probably on the way, so the wind will increase and rain is likely later—or snow in winter.

MAJOR CLUES **1**. Cirrus skies do not often look as organised as in **1**, but if the cirrus filaments mostly stream in the same direction, the weather may deteriorate rapidly. **2**. The crossed-winds rule for deterioration should be evident (surface wind blowing into the back of the observer while the Ci streams in from their left hand) showing that a warmer air mass is on the way, ie there must be a warm front or occlusion approaching. (For *left* read *right* in the Southern Hemisphere.) **3**. Good visibility (and heap clouds if present) show that you have a polar air mass and a temperature contrast with warm air to come, which often creates an active front. **4**. The appearance of ring haloes, as in **2**, show you now have the second in the sequence of clouds preceding an active warm front—cirrostratus—so it is more likely that wind and rain will follow.

EXPLANATION The model depression (Fig 2) leads to the expectation of a wedge of relatively warm air, extending hundreds of miles from where the warm front meets the surface. The cirrus of **1** forms in the high, thin, leading edge of the front. As the wedge deepens, the cirrostratus of **2** often covers most of the sky. Its appearance, with a halo following a build up of cirrus, makes it highly likely that wind and rain will follow. Darker lower layers of altostratus cloud follow, and then, eventually, low rain-bearing nimbostratus. In the Northern Hemisphere, winds to the south of the depression centre increase and back (shift anti-clockwise) as the front approaches whilst those to the north increase and veer (shift clockwise), although they may not always increase very markedly. In the Southern Hemisphere for *back* read *veer* and vice versa.

Element	Trend	Normal change	Normally in	Risk of	Possibly
Wind	Backing and increasing	Force 2–4 SW–S, becoming Force 5–7 S–SE	3–8 hours	Force 8–9	Force 4–5 in summer if depression weak
Visibility	Increase then decrease	10–13 mls or more, becoming 2–4 mls in rain, ½–1 ml snow	6–12 hours	Fog by end of period	4–6 mls in light rain
Precipitation	Rain or snow	Continuous slight to continuous moderate	6–12 hours	Heavy rain, thunder in summer half of year	Slight rain if depression weak
Cloud	Increasing to overcast	8/8 Cs then 8/8 As then 8/8 Ns with Sc and pannus	6–12 hours	8/8 low stratus covering coasts and hills	Broken skies if depression weak
Temperature	Cooler	Cooler as wind increases cooler still in pptn.	3–8 hours	Bitter in snow in winter	Little change if depression weak
Pressure	Falling	Moderate fall, followed by rapid fall later	10–20 hours	Major fall later with gales	Slow fall if depression weak

Warm front or occlusion approaching

MAJOR INFERENCE Moderate deterioration.
MAJOR CLUES **1**. Upper cloud (cirrus) and medium level cloud (altostratus) approaching. **2**. Upper wind strong and across trails as latter are being rapidly shredded. **3**. Lowest cloud (altostratus) aligned to show that upper cloud and wind is coming from the left of the lower (ie, from bottom left to upper right).

EXPLANATION The sky has the hallmarks of a warm front approaching, but there are breaks in the cloud build-up. This indicates an older and possibly less virile system than the skies of photographs **1** and **2**. Features in the As on the horizon are aligned in the same direction as those of the cirrus at the top of the picture. This means that the wind at As level is perpendicular to that at Ci level because Ci features stream with the wind, while As features lay across the wind (as with many stable clouds). Thus, the sky is set for deterioration but the breaks indicate a moderate and possible slow change to more wind and rain.

Element	Trend	Normal change	Normally in	Risk of	Possibly
Wind	Moderate increase and back	Force 5–6 SW to SE	6–12 hours	Force 6–7	Cyclonic Force 2–5
Visibility	Decrease later	2–4 miles in rain, ½–1 mile in snow	6–12 hours	½ mile in heavy rain 100 yards in snow	Fog over high ground later
Precipitation	Rain or snow	Continuous moderate	6–12 hours	Heavy rain over hills	Drizzle later
Cloud	Increase	8/8 As and Ns	3–6 hours	8/8 low St later	Broken cloud later
Temperature	Cooler	Cooler due to increase of wind	Next few hours	Cold in winter	Hot later in summer
Pressure	Falling	Moderate fall but if rapid go for stronger wind	8–12 hours	Rapid fall for gale	Slow fall for Force 2–3

SKY WHICH MEANS DETERIORATION

Altostratus ahead of a warm front or occlusion

MAJOR INFERENCE If this sky follows that in **1** with cirrostratus (haloes) between, then expect major deterioration. If it follows **2** and some Cs then expect deterioration. If following **3** often without Cs and with substantial breaks in the cloud build-up expect moderate deterioration.

MAJOR CLUES **1**. Organised flat and often featureless grey cloud through which the sun gradually recedes as if disappearing behind ground glass. **2**. Should have been preceded by Ci and Cs, but any cumulus will shred and die as the sun is cut off from the ground. **3**. Lineal features of As lying across the wind of the medium levels show the wind direction there and can be combined with the surface wind so that the Crossed Winds Rules can be applied.

EXPLANATION Altostratus like this when following Ci and Cs is an almost certain prognosis of rain and wind increase. Fragments of dark cloud which appear below the much higher altostratus layer can be the remains of cumulus and may indicate immediate rain or snow (see page 24). This will soon disappear as the higher cloud thickens. Steady thickening, lack of major features and a muddy greyness all contribute to a forecast of fairly immediate deterioration. Wind should be backing into southern quadrants and increasing when to south of advance of the depression. (For Southern Hemisphere see note on page 7.)

Element	Trend	Normal change	Normally in	Risk of	Possibly
Wind	Rapid increase and back	Force 6–7 in winter, Force 5–6 in summer	4–6 hours	Force 8–10	Cyclonic force 4–6 if shallow depression
Visibility	Increase then decrease	Tens of miles, then 2–4 miles in precipitation	2–4 hours	½ mile in snow (or less)	Generally good
Precipitation	Rain or snow	Continuous moderate	2–4 hours	Heavy rain or snow	Intermittent, slight pptn.
Cloud	Increase and lowering	8/8 Ns with pannus **5** and virga **6**	2–4 hours	Fog over hills and at sea	Broken Sc and St later
Temperature	Cooler	Cooler due to wind increase and no sun	Next hour or two	Very cold in winter	Warmer later if cloud breaks
Pressure	Falling	Moderate or rapid fall	4–6 hours	Major fall for severe gale	Slow fall if shallow depression

SKY WHICH MEANS DETERIORATION

5 Imminent rain or snow

MAJOR INFERENCE Rain or snow will commence very shortly and sometimes almost immediately.
MAJOR CLUES **1**. Totally overcast, lowering sky coupled with— **2**. Ragged, developing lumps of grey cloud coming from windward below the main cloudbase.

EXPLANATION This sky says 'Reach for the mac, oilskins or umbrella'. There are all the signs of an approaching warm front. The sun has long since disappeared behind the grey featureless altostratus. The camera was looking into the south and the pannus is forming in the wet air below the base of the As and streaming in on the lower wind. Such cloud indicates that precipitation is already falling—although it may not have yet reached the ground. This sky may look like the previous one, but whereas the low cloud is dispersing in 4, in 5 it is increasing.

Element	Trend	Normal change	Normally in	Risk of	Possibly
Wind	Increase	Increase by 1–2 Beaufort Forces	A few minutes	Gusts	Little change
Visibility	Decrease	Lower by a few miles in rain to less than 1 mile in snow	Half an hour or an hour	Fog limits in snow	Little change
Precipitation	Immediate commencement	Intermittent at first, becoming continuous moderate	A few minutes		Wholly intermittent
Cloud	To total cover of pannus	6–8/8 pannus or Sc and St below 8/8 As and Ns	An hour or two	Covering hills	3–5/8 pannus or Sc
Temperature	Immediate fall	Cooler in downdraughts caused by falling rain	Tens of minutes	Suddenly very cold in winter	Little immediate change if rain slight
Pressure	Continuing fall	Moderate or rapid fall	2–6 hours	Major fall	Slow fall if shallow depression

24

A front passes

MAJOR INFERENCE Change imminent. In bad weather the break is often not as marked as here. It may just be a lightening of the sky behind a trailing cloud base. Expect gusts and veering wind, a steadying off or rise in pressure, a change of air mass and change of humidity and air temperature, and a change in degree of precipitation or in its form.

MAJOR CLUES **1**. This is a warm front when total low cloud cover and cool rain changes to a higher cloudbase and cessation of rain. Lower cloud may, however, soon close in again with high humidity and drizzle. **2**. This is a cold front when warm, humid tropical air changes to cool, moist polar air and poor visibility changes to good visibility. There is a period of rain and showers. Cumulus cloud can be seen on the windward horizon. **3**. This is an occluded front when the warm front is followed by the cold front without any tropical air mass between.

EXPLANATION A photograph of a weak warm front passing was chosen because more virile fronts show few contrasts in their features. However, the trends are illustrated. There is a low cloud line with trailing virga. Beyond there is obviously a change of weather type. Rain now falling can be expected to cease as the lowest cloud passes.

Element	Trend	Normal change	Normally in	Risk of	Possibly
Wind	Immediate veer, gusty	Veer behind lowest cloud with wind increase	Tens of minutes	Heavy gusts from left hand of present wind	Wind decrease or little change of direction
Visibility	Decrease	Decrease as front passes. Poor behind warm front. Increase behind cold front	Tens of minutes	Fog behind warm front	Little change when front weak
Precipitation	Immediate cessation of continuous moderate rain	Very low cloud and drizzle when a warm front. Showers after a time behind cold front	Tens of minutes to an hour or two	Heavy as front clears	None if front weak
Cloud	Lowering, lifting, then lowering again	8/8 Sc and St if warm front, 4/8–6/8 Cu of Cb behind cold front	Half an hour or hour	8/8 low stratus behind warm front	Broken skies with Sc when front weak
Temperature	Warmer if warm front. Cooler if cold front	Muggy due to increased humidity and air mass change when warm front. Fresher when cold front	Tens of minutes		Hot in summer under broken skies
Pressure	Falling, then steady or rising	Steady off if warm front. Rise if cold front	Half an hour or hour	Further fall behind warm front if depression vigorous. Sharp rise behind cold front	Little change with weak fronts

Thundery sky

MAJOR INFERENCE Thundery outbreaks within the next 12 hours or less. Use of the Crossed Winds Rules will tell if an unstable warm front is downwind and so warn of thunderstorms. However, see 17.

MAJOR CLUES 1. Undulating lines of altocumulus castellanus, accompanied by generally disorganised groups of altocumulus floccus with filaments, streaks or small dapples of cloud showing dark against higher clouds. 2. General regime preceding thundery weather. High temperatures, little wind, possibly high humidity.

EXPLANATION This is the classic 'thundery sky'. The cloud in the lower centre is Ac Cast, which always indicates instability in the middle layers of the atmosphere and so strong risk of thunder later. Such skies often occur in the forenoon and are followed by clearer skies in the afternoon which allow the land to heat and so prepare the whole lower atmosphere for thunderstorms by evening. This sky, with the orientation for deterioration given in the Crossed Winds Rules, provides a confident forecast of thundery outbreaks.

Element	Trend	Normal change	Normally in	Risk of	Possibly
Wind	Moderate increase	Force 3–4, highly variable	12 hours	Force 7 gusts. Possibly Force 9 gusts	Tornadoes or waterspouts later
Visibility	Decrease	3–6 miles	12 hours	Fog on coasts	Little change
Precipitation	Heavy showers, lightning and thunder	Intermittent or sporadic heavy rain plus hail or rain showers	6–18 hours	Heavy hail later	Little precipitation
Cloud	Decrease, then increase	6–8/8 chaotic As and Ac with Cb and Fr St	6–18 hours	Heavy frontal thunderstorms	Clearance without major deterioration
Temperature	Very hot, followed by *catastrophic fall* in thunderstorms	Cooler	6–18 hours		Remaining warm
Pressure	Slow fall	Not very marked fall	6–12 hours	Heavy fall with frontal thunderstorms	Little change

Thunderstorm

MAJOR INFERENCE Imminent gusts (up to 40–60 knots at their worst), large drop in temperature, thunder and lightning, hail and heavy rain.
MAJOR CLUES 1. Characteristic anvil tops to the Cb clouds
2. Violent upward motions or rolling motions can often be seen in the nearer cloud arches with curious wave-like clouds arriving just ahead of the worst of the wind.

EXPLANATION The vertical extent of thunder clouds which threaten is often impossible to assess because of surrounding cloud. This photograph shows the rolling cloud ahead of an imminent storm. Note the 'rain curtains' in the lower middle of the scene.

Element	Trend	Normal change	Normally in	Risk of	Possibly
Wind	Violent increase, direction change	Often goes from 5 kt towards storm to gusty Force 5–6 from storm	Half an hour or less, depending on proximity	Gusts to 40–60 kt	Only Force 3–4, gusty
Visibility	Sudden decrease in rain and hail	½ mile or less in heavy rain or hail	Variable time as rain also occurs intermittently, violently	100 yards or less in heavy hail	
Precipitation	Heavy rain or hail showers, thunder, lightning	Intermittent heavy and light periods of rain or hail	Periods occupying approx 10 minutes light and heavy rain, light in rear of storm	Heavy and large hail in worst storms (rare)	Cloudburst, especially in hilly areas late in the day
Cloud	8/8 Cb and related cloud	Rapid increase and very dark waved and rolled clouds under leading edge	5–10 minutes from cloud edge overhead	Tornadoes or waterspouts	Big Cu and As on edges of main cells
Temperature	Dramatic fall	From hot and oppressive to cold, with wind and rain	A few minutes from leading edge of cloud		Little change if storm is just passing
Pressure	Falling slowly	Minor variations under thunderstorms	A few hours	Rapid fall with worst storms	Rising slowly

Warm sector weather

MAJOR INFERENCE Before there can be a major change of weather type a cold front must pass—or several minor fronts—so expect continuation of the weather pattern associated with tropical maritime airstreams. Keep an eye on the sky for major change coming from windward. Expect a lot of cloud but if it clears then consider risk of low stratus at night or sea fog over sea and coast.

MAJOR CLUES **1**. Predominantly cloudy airstream. **2**. Sometimes poor visibility. **3**. Warmth and humidity (often muggy), some drizzle or orographic rain.

EXPLANATION Warm sector weather comes in many varieties depending on how pure it is. It follows passage of a warm front and may last for a few hours or as many days. Immediately behind fronts it is normally totally cloudy, prone to drizzle and fog especially over hills and coasts. Clouds are most often those of stability, ie, stratus and stratocumulus plus cloudy islands in the medium and high levels. However, with bright conditions in the summer, cumulus clouds can form over land or, as here, the sea may be warmer than the air and set off heap clouds. Lens-shaped altocumulus lenticularis forms in the lee of hills, as well as cirro and alto forms of this cloud shape. Sometimes, especially in summer, warm sectors can be thundery.

Element	Trend	Normal change	Normally	Risk of	Possibly
Wind	Remain as now	Diurnal variation of wind speed		Increase to Force 5–6	Decrease
Visibility	Poor	Fog and low cloud	During overnight hours	50 yards or less, especially in winter	Little change
Precipitation	Drizzle on hills and coast	Dry-out with time	Over a day or so	Rain, thunder etc later	Very humid
Cloud	Cloudy	4–7/8 low cloud with patchy medium and high cloud	Usually most cloudy by night and least in latter part of day	8/8 Stratus at 100 ft or less	No cloud
Temperature	Warm for time of year	Follow diurnal variation	Maximum temperature after lunch. Minimum after dawn	Summer temperatures in the eighties, cool under stratus and in fog	
Pressure	Little change	Fall on approach of cold front	In days or hours	Rapid fall and sudden change of air mass, with thunder	Rising steadily

MAJOR INFERENCE Fog-banks form over the sea when there is a tropical maritime airstream. In any case the air is warmer than the sea over which it finds itself. Often quite unpredictable in occurrence and extent. Clears as quickly as it arrives. Warning of sea fog sometimes comes from hearing fog signals at sea. In spring and summer, sea breezes may bring it in over beaches. A hot day inland may find coastal resorts wreathed in cold, clammy fog. Also presages formation of low stratus over land and sea at night.

MAJOR CLUES **1**. Warm and humid airstream probably warmer than the sea. **2**. Wind or sea breeze to bring fog in off the sea. **3**. Fog signals heard.

EXPLANATION The coast is often a zone of transition for foggy conditions. Warm, moist onshore winds bring in the fog which soon disperses as it arrives over warm land areas. Shallow sea fogs can clear along the coast due to convection currents formed over coastal slopes. Coastal sailors can frequently foretell imminent sea fog by seeing it forming over coasts before it spreads back out to sea. This fog is not the kind which forms over land during the night and disappears by day. Sea fog can persist for days in some cases.

Element	Trend	Normal change	Normally	Chance of	Possibly
Wind	To come off the sea	Light onshore when morning is sunny	Occurs in the hours before and around lunchtime	Complete calm	Sea fog near fronts, with Force 4–7 winds
Visibility	Fog or mist	50–100 yards over coastal areas, less in banks over sea	Unpredictable, but fog-banks drawn in by sea breeze during the day	10–20 yards	No fog, but generally hazy visibility
Precipitation	Little or none	Drizzle where fog-banks lift over coast and hills	*As for visibility*	Light rain	No precipitation at all
Cloud	Half to full cover	Obscured sky when fog-banks invade	*As for visibility*	Total cover at sea level	No cloud
Temperature	Cool and clammy in fog	From warm and humid to cool and humid	In next few minutes	Cold near the coast all day	Remaining warm
Pressure	Usually no appreciable change				

Showers

MAJOR INFERENCE   Showers in the next hour or two. They may not be heavy but can be expected to become heavy later. General cloud cover of 4–6/8 indicates a sprinkling of shower clouds over level terrain. Showers will grow preferentially over hills and die out in their lee. If showers are indicated, but do not develop, then possibly—(a) an air mass trough is upwind, as in **12**, or (b) the upper air is subsiding. Look for the cirrus of another depression.

MAJOR CLUES   **1**. Big cumulus, ie, distance from the base to the tops is greater than the distance from the base to the ground. **2**. Developing solid heads—no spindly chimneys. **3**. No cloud layer above to inhibit upward growth. **4**. Cool polar airstream with wind typically Force 3–4.

EXPLANATION   The prototype of showery airstreams is the unstable northwesterly which arrives behind major depressions. Its showers form primarily on windward facing coasts and the hills inland from them. The shower activity falls away the further one goes from the windward coast, but showers frequently sweep right across such comparatively narrow lands as Britain. Early growth of ragged Cu (ie, by 0900–1000 GMT) and a cool blustery wind are signs of a showery day to come.

Element	Trend	Normal change	Normally	Risk of	Possibly
Wind	Normally NW or W, marked diurnal variation	Gusty in afternoon, less at night	Diurnally	Force 6–7	Force 2–3 when showers are heavy with thunder
Visibility	Very good	Very good by day, except in showers, bright stars at night	Diurnally	Dawn fog in sheltered industrial areas	Good all night if wind stays up
Precipitation	Rain, snow, occasional hail	Increase with day, die out in the evening	Diurnally	Occasional thunder, especially over hills	Continuous showers over the sea
Cloud	6–7/8 Cb and Cu	No early cloud, maximum shower cloud in middle of day, dying out with evening	Diurnally	Thunderstorms	6–8/8 Sc when air is stabilising
Temperature	Cool	Cold at night, cool during the day	Diurnally	Frost in sheltered spots at night	
Pressure	Steady or rising slowly	High pressure	Over days	Rapid rise and subsequent fall	Slow fall

MAJOR INFERENCE Showers and some more continuous rain, possibly some thunder. Expect the whole thing to be over in an hour or so but do not expect a major change of weather type. Cool showery weather is to be expected after the trough, as it is before it. It is not like a front which separates different air masses.

MAJOR CLUES **1.** Little cloud ahead of trough in clear polar airstream. **2.** Obvious line of shower and other clouds moving from windward. **3.** Some backing in the wind direction (veering in Southern Hemisphere).

EXPLANATION The cool polar maritime airstream breeds air mass troughs. Sinking air ahead of (and behind) the trough leads to fewer showers than would be expected and may lead to suspicion that an as yet invisible trough is somewhere upwind. Little obvious trough activity occurs on windward facing coasts, but it builds up as one gets further inland. Major thunderstorms are not to be expected but waterspouts may be produced over the sea and dust-devils over land.

Element	Trend	Normal change	Normally in	Risk of	Possibly
Wind	Force 3–5, westerly	Squally, Force 4–6 W–NW	Half an hour to an hour	Gusts of Force 7 or 8	Veering Force 3–5 NW
Visibility	Very good	1–2 mls in showers	Half an hour to an hour	Less than ½ mile in snow	Very good
Precipitation	Fair or fine	Showery, rain or snow	Half an hour to an hour	Thunder	Fair or fine
Cloud	2–4/8 Cu, 3,000–5,000 ft	7–8/8 Cu, Cb and As. Lowest base 1,000–2,000 ft	Next half hour	Temporarily very low base in snow	2–6/8 Cu or Cb
Temperature	Cool for time of year	Temporarily colder	Next half hour		Cool again
Pressure	Falling slightly	Small fall	Half an hour to an hour	Sharp fall in worst cases	Slow rise

SKY ASSOCIATED WITH SUDDEN LOCAL CHANGE

MAJOR INFERENCE Showers and/or thunderstorms with heavy rain and perhaps hail. A major storm area possibly lies hidden behind the more obvious clouds.

MAJOR CLUES **1**. Large, solid cloud tops growing upward, but separated (compare **8**). **2**. Much surrounding cloud, which is relatively low (compare **12**). **3**. Other signs of a polar airstream.

EXPLANATION Individual cumulonimbus clouds may exist, as here, over summer cornfields, or over winter snow cover, or wedged in between smaller cumuliform clouds in cool, unstable airstreams at any time of year. They may blow by in well-washed airstreams, give a couple of claps of thunder and be gone. In summer the situation depicted may presage the development of a more major storm area, but in this particular case there is little of the attendant dark cloud masses which go with big thunderstorms. The weather type differs from that of **8** by being relatively cool and not sultry. Here the cloud cells are well separated from their neighbours so that showers are also well separated and any thunder that may occur will be relatively insignificant.

Element	Trend	Normal change	Normally over in	Risk of	Possibly
Wind	Light wind possibly toward storm in worst cases	Squally, gusts to 30–40 knots in worst cases	½–2 hours	Tornadoes or waterspouts	Force 2–4 from another direction
Visibility	Moderate	Temporarily poor in precipitation	½–2 hours	Less than ½ mile in snow showers	Moderate to good
Precipitation	Fair or fine	Heavy showers of rain or snow	½–2 hours	Hail and thunder	Fair or fine
Cloud	2–6/8 Ac and As, Cu	6–8/8 Cu and Cb with As and Ci		Funnel clouds	2–6/8 Cu
Temperature	Cool	Cold in showers	½–1 hour		Cool
Pressure	No appreciable change			Moderate fall	

Quiet evening

MAJOR INFERENCE A calm night with fog and mist patches especially in industrial areas and in hollows, river valleys etc. Possibly air and ground frost.

MAJOR CLUES 1. Lack of wind both at the surface and aloft. 2. Strongly stabilising air near the ground causing formation of mist and smoke decks. 3. Little or no low or medium cloud to prevent radiation from the ground.

EXPLANATION The smoke deck shows how rapidly the earth is cooling and so stratifying the air layers over it. The smoke is trapped in an 'inversion' where cool air changes to warmer air above it. This inversion layer will move upward with the night and trap water vapour and smoke near the ground so leading to mist and fog. If the ground falls below freezing then the fog risk is less as some of the moisture is now held in suspension as solid ice crystals on grass etc.

The upper cloud is largely spread-out vapour trails, but they are dense and persistent. There seems, however, to be no organised wind direction at their level and this points to a continuation of the present quiet weather.

Element	Trend	Normal change	Normally over by	Risk of	Possibly
Wind	Calm	Very light offshore land breezes near coasts, especially in the autumn	Midnight	Total calm	Wind picking up before dawn
Visibility	Decrease	Variable mist and fog	Midnight or early hours	Less than 20 yards	Increasing if wind picks up
Precipitation	No true precipitation	Dew and hoar frost	Early hours	Icy patches on roads	No frost
Cloud	Dispersing	Clear skies	Evening	Increase with the night	8/8 low St, 4–6/8 Sc or increasing Ci
Temperature	Falling	Fall to ground frost and air frost in winter half of year	Dawn or before	Air frost in winter	No frost even in winter if wind and cloud increase
Pressure	Generally steady, possible slight changes, rising or falling			Fall later in night if depression approaching	

SKY ASSOCIATED WITH IMPROVEMENT

MAJOR INFERENCE The night will be clear and the temperature will drop. The fractostratus clouds below the main cloud base indicate that the air there is still wet from the rain falling out of the cold-frontal clouds above. This wetness left on the ground in the evening and the fall of wind speed with the night may well produce mist and fog patches in areas sheltered from the wind. Frost can also occur in the winter half of the year.

MAJOR CLUES **1**. Good expanse of upper cloud on which the sun can project red rays. **2**. Immediate past history of cold front or occlusion (rain, showers, low cloud etc). **3**. Obviously well-broken skies behind the front.

EXPLANATION The low sun of sunset (or sunrise) always projects red rays. These do not of themselves presage a fine night. The fine night follows because the sun can only illuminate the windward clouds if there is no cloud cover over the horizon (ie, to windward at sunset). The clouds in the foreground must be at medium and high levels. These conditions are fulfilled when a cold front has just passed and polar air is moving over the observer. It will then be a fine night.

Reverse the situation and you have the dawn of a bad day. The sun is now to leeward and shining over a clear horizon onto the clouds of a warm front whose weather and wind has yet to come and is that of which shepherds warn.

Element	Trend	Normal change	Normally over by	Chance of	Possibly
Wind	Decrease	W to NW, Force 2–3. Calm in sheltered areas	Early hours of morning	Calm by dawn	Increasing again by dawn
Visibility	Decrease	Mist and fog in sheltered surrounding areas	Early hours of morning	Less than 50 yards in smoke	Clear all night if wind stays up
Precipitation	None, but perhaps ground frost	Hoar frost by dawn in winter half of year	Dawn	Air frost in winter and spring	None if wind stays up
Cloud	Decrease	Clear skies	Midnight	8/8 low cloud when wind stays up and ground is left moist	Cloudy aloft by dawn if another front on the way
Temperature	Falling sharply	Cool but clear	Dawn	Very low in winter	Little change if wind stays up
Pressure	Rising	Steady rise		Sharp rise and signs of fall by dawn	Little change

MAJOR INFERENCE The centre of a depression has passed eastwards and so the weather should improve with the wind decreasing, possibly only slowly, while pressure should be rising, possibly rapidly. Risk of showers by day, but there will be clear skies by night over land. Possibly showers day and night over the sea.

MAJOR CLUES **1.** Crossed Winds Rules applied to upper wind and lower wind indicate that cold air is moving in, ie, polar air behind the cold front of a depression. **2.** Otherwise the sight of blue sky shows polar air which normally arrives behind a receding depression.

EXPLANATION Here the cumulus is riding in the north-west wind behind the cold front of a passing depression. There is what looks like jet cirrus above – which it actually is – but the cirrus is moving from south-west – not north-west as in the photo on page 16. Applying the Crossed Winds Rules indicates that the approaching clear weather will persist for a while and not swing back into clouds and rain. There are still likely to be showers however.

Element	Trend	Normal change	Normally over in	Risk of	Possibly
Wind	Slow decrease	Force 3–5 by day. Force 2–4 at night	One day	Increase later when pressure rises rapidly	Squally in showers
Visibility	No great change	Obeys diurnal variation	Half a day	Fog and mist by night in sheltered industrial areas	Very good
Precipitation	Showers by day	Clear by night over land. Showers both day and night over the sea		Frost at night in winter half of year	No showers
Cloud	Decrease	Diurnal variation of Ci or Cb cloud	Half a day	Increase of cirrus later if another front on the way	
Temperature	Cooler	Diurnal variation over land		Freezing	
Pressure	Rising	Rising slowly for real improvement	A day or two	Rapid rise and later rapid fall	Little change if depression weak

Will it thunder?

MAJOR INFERENCE It looks as if thunder is possible because of the way the darker clouds are broken into globules, and are lying along the wind at their level. The layer of small altocumulus above is typical of these kinds of situations. However, it is not likely to develop into thunder because the wind off the sea is roughly in the direction from which the clouds are coming (see Crossed Winds Rules). On the south coast of England, where this photograph was taken, thundery-looking outbreaks often come from France but they do not always develop into thunderstorms.

MAJOR CLUES **1**. Minor cloud elements broken into a 'thundery sky', but major features streaming parallel to surface wind indicate no front downwind. **2**. Little or no lower cloud—there is much alto cloud in these situations.

EXPLANATION This is a thundery sky just as **7** is. However, a thundery sky takes many forms and this one is different. The sort of clouds depicted here, and also in **7**, run ahead of thunderstorms in the medium levels, but there is a difference in timing. The clouds of **7** are formed often many hours ahead of the thundery outbreaks, while the cloud in this photograph is likely to be part of the immediate forerunning cloud.

Element	Trend	Normal change	Normally over	Risk of	Possibly
Wind	Will remain in same direction, but increase a little	Force 2–4, gusty and perhaps variable	In a few hours	Thunder if surface wind backs	Passing, and later fine
Visibility	No great change	3–6 miles by dawn	During night	1–2 miles in places	No change if wind increases
Precipitation	Little or none	Few thunder spots, if any	During next few hours	Thunderstorms at medium level if wind backs	None at all
Cloud	Not much change	6–7/8 thundery Ac, and As with Cc and dense Ci	In a few hours	8/8 thundery Ac if wind backs	Clearing to 2–4/8 Ac and As
Temperature	Falling	Slight fall			
Pressure	Slight fall	Fall by less than 1 mb in 3 hours	During next few hours	Falling again later, hence deterioration with risk of thunder	Rising, and so major improvement

Will it rain?

MAJOR INFERENCE   It is very unlikely to rain more than a few small spots. This is because the cloud base has hard, downward protuberances. Being hard they indicate a dry cloud base and the downward lumps indicate sinking currents. Both are against rain or showers.

MAJOR CLUES   **1.** Big cloud area which might produce rain but whose base is hard and lumpy. **2.** No real sign of any amorphous areas from which rain might be falling. No pannus, no fractostratus. **3.** No sign of reduced visibility in any direction, which might indicate falling rain.

EXPLANATION   Black-looking cloud areas usually mean rain. When udder-like protuberances (mamma) can be seen then this means little rain (or very weak showers). They can often be seen in the rear of big cumulonimbus clouds indicating that the worst is over—at least from *that* cloud. Sometimes the mamma is more striking than here, as when an old front is passing. Cloud areas take a long time to disperse when they were once thick and this so-called 'frontolysis' leads to large cloud sheets from which no precipitation falls. Sometimes, however, an old front may rejuvenate and rain break out when none was expected.

Element	Trend	Normal change	Normally over	Risk of	Possibly
Wind	To veer and increase slightly	Veer behind the front; strongly stabilising air (Sc) indicates no great increase in speed	Within next half an hour	A few gusts	In areas prone to them pronounced mamma may indicate tornadoes
Visibility	Increasing	5–10 miles	Next hour or so	Decrease after nightfall	
Precipitation	Little or none	A light shower, possibly just a few spots of rain	Next half an hour	A heavier shower if whole cloud base is not hard	No rain at all, but take note of above remarks concerning tornadoes
Cloud	Decrease	Fair or fine	An hour or so	4/8 Cu or Cb during afternoon if front clears in forenoon	7/8–8/8 Sc later
Temperature	Little change	Diurnal variation			Cooler
Pressure	Rising	Rising slowly	Next 6 or more hours	Falling later if another depression is on the way	No change

SKY ASSOCIATED WITH NO IMMEDIATE CHANGE

MAJOR INFERENCE Fair weather and light to moderate winds. Sea breezes developing with the day near coasts in spring and summer and, perhaps, penetrating tens of miles inland, as well as the same distances out to sea.

MAJOR CLUES **1.** Cumulus clouds that have flat bases and subdued tops. **2.** Distance from base to top less than distance from base to surface. **3.** Ranged in ranks or streets. **4.** Clear area to seaward indicates sea breeze over this area.

EXPLANATION We are inland looking towards the sea, which is about 10 miles away. Looking under the line of cloud in the background we see a clear slot with no clouds. That line is a seabreeze front moving in from the coast—the mechanism of the breeze has led to air sinking over the coast and eroding the clouds there, leaving it sunny. It will soon be completely sunny here as well.

NB On good sea breeze days (typified by early morning wind light to moderate, partly unstable airstream with half cover of Cu or less) the onshore sea breeze cannot be used with the Crossed Winds Rules.

Element	Trend	Normal change	Normally	Chance of	Possibly
Wind	Diurnal variation	Force 2–3 by day. Force 1–2 by night	In a day	Sea breeze	Force 4–5 on the edges of anticyclones
Visibility	Good	Follow diurnal variation, but rarely less than several miles	In a day	Haze inland	Poor locally at night in the winter half of year
Precipitation	None			Showers over hills	
Cloud	Diurnal variation	3–6/8 small Cu by day, little cloud at night	In a day	6/8 Cu along sea breeze fronts	No cloud over coasts
Temperature	Diurnal variation	Maximum by early afternoon. Minimum by dawn	In a day		
Pressure	Steady or rising	Slow rise	In hours or days	Falling when low is encroaching	

Cirrus revealing no change

MAJOR INFERENCE The situation will remain largely static, with no marked deterioration or improvement, because the upper wind (revealed by the trailing fallstreaks) and the lower wind (indicated by the way the low clouds lie in lines more-or-less parallel with the fallstreaks) are shown to be in roughly the same direction. If there is to be a marked change of air mass then the upper and lower winds must be crossing one another at a large angle.

MAJOR CLUES **1.** Cirrus which might mean the onset of a warm front, but the parallel winds tell of no great change. **2.** No other obvious signs of trouble, such as tumbling barometer.

EXPLANATION A sky with cirrus in it may mean many things. Photographs 1, 2 and 16 show how to sort out some of the messages. There is undoubtedly some wind aloft and that might mean some increase at the surface later. No warm front is indicated downwind so whatever happens it cannot be very bad. Care must be taken to monitor the surface wind direction because on the back of a ridge of high pressure upper and lower winds can be parallel. As the ridge moves over, however, the surface wind backs and so becomes crossed to the upper wind, which probably will not change its direction much. Now deterioration is on the way!

Element	Trend	Normal change	Normally over	Risk of	Possibly
Wind	Little change	Force 2–4 by day over land	In some hours	Force 4–7 if surface wind backs later	Little change
Visibility	Little change	10 miles or more as wind freshens	In some hours	Deterioration at night	Little change
Precipitation	None				
Cloud	Increase aloft	3–6/8 Ci and Cs with some Ac and As	In some hours	Total cover of high and medium cloud	No great increase
Temperature	Slight decrease	Cooler due to wind increase	Soon	Colder in winter	Little change
Pressure	Little change				

MAJOR INFERENCE No change of weather type yet. Sometimes Sc may run ahead of bad weather, but not often. Stratocumulus in the centre of winter anticyclones can accompany the worst of fogs and smogs, so in this case it goes hand-in-hand with a very bad weather type. Yet it is a cloud of a largely unchanging situation. Winds are normally light to moderate and shift direction, but not very rapidly.

MAJOR CLUES **1.** Flattened cumuliform elements in large pieces with some lighter chinks between the elements. **2.** Airstream generally stable, ie, no real shower clouds (Cu or Cb). **3.** Often accompanies tropical airstreams which have dried out. **4.** Does not normally attend a bad weather system except on its periphery.

EXPLANATION Here on the South Downs in Sussex, where the Roman roads meet at Bignor, the setting sun colours a great sheet of strato-cumulus. Stratocumulus forms under inversions. It may have been cumulus which has tried to grow through an inversion and has only succeeded in flattening sideways under it. This is the way clear mornings sometimes become totally cloudy. Such inversions are relatively strong, but in a benign mood, as depicted here, Sc is a normal accompaniment to a modified polar airstream which has travelled a long way and in which air from aloft has sub-sided to produce the inversion.

In association with nimbostratus in bad weather Sc may form extensively below the main cloud base so that three cloud decks appear; a low cover of pannus **5** topped by a total cover of Sc above which hides the really thick rain-bearing Ns clouds.

Element	Trend	Normal change	Normally	Chance of	Possibly
Wind	Little change	Force 2–4 by day. Force 1–3 by night	Diurnally	Calm	Picking up from new direction if a low is encroaching
Visibility	Decrease	5–8 miles	Diurnally	10 yards in smoke in winter half of year	Smog in industrial areas in winter
Precipitation	None				
Cloud	Variable	6–8/8 Sc sometimes, but may break to 1–3/8 at times	In hours	Total cover in winter with fog	Variable, with clearances and then clouding again
Temperature	Little change	Remain relatively warm		Cool in fog	Cold in winter
Pressure	Little change	Slow rise or fall	In days	Sharp fall if depression threatens	

MAJOR INFERENCE A warm moist tropical maritime air mass. With a relatively strong wind low stratus comes in the rear of active warm fronts, ie in the warm sectors of young and active depressions. If you see it forming in the evening expect total cover as night begins to fall. Bad for aircraft movements and mountaineering.

MAJOR CLUES **1**. Undefined and amorphous base—often covering whole sky. **2**. Airstream moist and muggy. **3**. Accompanied by flurries of drizzle when at its wettest, but not normally by rain, except on windward-facing hill slopes.

EXPLANATION A grey and featureless day. Stratus is low and formless—it is fog raised above the earth's surface. It also forms in the easterly winds to the north of depression centres and may obscure the sky on days when thunderstorms are going to erupt later. It must have rain or drizzle falling into its airspace to keep it from breaking up when winds are strong to gale. Overland it may be lifted by the wind into lumpy turbulence cloud.

Element	Trend	Normal change	Normally in	Risk of	Possibly
Wind	Little change for hours at least	Veer and gust at cold frontal passage	Hours or days	Failing light, when stratus will become fog	Dispersing if wind increases
Visibility	Poor but variable	½–2 miles	Periods of hours or minutes	50 yards if wind lessens	
Precipitation	Drizzle, rain over hills	Unpredictable drizzle patches especially near windward coasts, continuous rain over hills	Tens of minutes to some hours	Rain and showers if near depression centre	No drizzle or rain in less wet airstreams
Cloud	Overcast	8/8 St at 100–300 ft, 8/8 Sc or Ns above	Unpredictable times	Sky obscured and thick fog	2–6/8 Sc or other low cloud if wind increases
Temperature	Muggy	Warm for time of year if cloud breaks	Unpredictable times	Cool or cold under total cover	Hot or warm if cloud breaks
Pressure	Steady or falling	Moderate or slow fall then rising	Hours or days	Falling rapidly if depression centre not yet passed	Rising if centre has passed

Altocumulus and cirrocumulus

MAJOR INFERENCE Cirrocumulus and altocumulus sheets indicate a generally unstable tendency associated with air layers which are being lifted. They often accompany some high level frontal activity, or they may be due to the general edging up of the high layers when a heated land surface forces the lower layers to expand. 'Mackerel sky, mackerel sky. Not long wet, and not long dry' may accurately sum up these clouds.

MAJOR CLUES **1**. Islands, areas, patches of closely-packed globular elements. **2**. Cirrocumulus has no shadows, altocumulus has light shading. **3**. Often Ac shows coloured patches when illuminated by sun or moon. **4**. Increasing in amount for possible deterioration, decreasing for probable improvement.

EXPLANATION Unstable cloud elements at high and medium levels when orientated for deterioration across the surface wind can mean that rain will be locally heavy. In this photograph the Crossed Winds Rules indicate that there is no great change to be expected for some hours at least, because the surface wind is parallel to the major cloud features aloft.

Element	Trend	Normal change	Normally	Risk of	Possibly
Wind	Increase later	Increase somewhat when preceding fronts	In hours	Major increase later	Little change
Visibility	Moderate or good		Diurnal variation		
Precipitation	Intermittent later	Rain, with a thundery tendency later	In hours or may be a day or more	Heavy rain if wind backs	Remaining fair
Cloud	2–6/8 Ac and Cc	3–6/8 Ac and As, with Ci and Cc	In next few hours	8/8 thundery As and Ac later	Cloud decreasing
Temperature	No great change	Diurnal variation	Diurnal variation but cooler if rain comes	Cold in thundery rain	Becoming warm
Pressure	No great change	Fall or rise by small amounts	In hours or days	More rapid fall later if depression approaching	Rising for better weather

Coastline clouds

MAJOR INFERENCE Looking in the direction of the sea from a few miles inland the observer sees an unstable airstream over the water only. This indicates that the sea is warmer than the land and is a phenomenon primarily of autumn and winter. If the wind shifts or the warming up of the land brings a sea breeze (least likely in late autumn) then showers could invade the land. Expect gusty showers if leaving harbour.

MAJOR CLUES 1. Definite and obvious change of cloud condition near the coastline. 2. Wind blowing mainly parallel to or off the coast in a showery situation; but blowing onshore without cloud over coast or sea in sea breeze, see 19. 3. Knowledge of probable sea temperature and air temperature helps to decide likely conditions outside.

EXPLANATION The yachtsman can assess the conditions outside harbour from just inland. Winds increase by 1–2 Beaufort Forces, or more, when they leave the shelter of land. Take maximum gust speed in harbour to be representative of the mean wind at sea and reef accordingly.

A coastline may form a zone of change as follows: 1. Air unstable over the sea, but stable over the land (as shown here). 2. Air stable over the sea, but unstable over the land. If the wind is parallel to the coast, or blowing offshore, then the reverse of the depicted situation can occur with cloud over the land but none over the sea.

Element	Trend	Normal change is to	Normal times	Risk of	Possibly
Wind	Increase, gusty at sea	Force 3–4 over the land to 4–5 over the sea	On reaching water with open sea fetch	Force 6, with gusts to 7 or 8	Force 2–3 over land and 3–4 over the sea
Visibility	Poor in showers at sea	6–10 miles out of showers to ½–3 miles in showers	Showers arrive and pass in minutes	Less than ½ mile in snow showers	Over 30 miles
Precipitation	Showers of rain or snow over sea	Infrequent showers inland and frequent showers over sea and immediate coastal strip	On leaving coastal strip for sea	Hail and thunder, but passing	Frequent showers over land and few over sea
Cloud	2–4/8 big Cu and Cb over land, 4–7/8 Cu and Cb at sea	Seaborne cloud over land when wind shifts off sea, land distribution of cloud found off coast when wind shifts off land	Several hours or more, perhaps rapidly if land strongly heated	Funnel clouds, ie waterspouts over sea	
Temperature	Feels cooler over the sea	From normal for time of year on land to cool in showers over sea, despite higher sea temperature	On entering showery area		

Index